Mathematical Vocabulary Book
INTRODUCTION

Who is this book for?

The purpose of this book is to identify the words and phrases that children need to understand and use if they are to make good progress in mathematics. It is designed to support the National Numeracy Strategy alongside the *Framework for Teaching Mathematics*.

This booklet will be of particular interest to you if you are:

a class teacher

a member of staff supporting pupils learning English as an additional language

a special needs teacher or assistant

a classroom assistant working with pupils in mathematics lessons

a parent or other adult supporting children in class or at home

Why is the book needed?

There are three main ways in which children's failure to understand mathematical vocabulary may show itself: children do not respond to questions in lessons, they cannot do a task they are set and/or they do poorly in tests.

Their lack of response may be because:

they do not understand the spoken or written instructions,
such as 'draw a line between...', 'ring...' or 'find two different ways to...'

they are not familiar with the mathematical vocabulary,
that is, words such as 'difference', 'subtract', 'divide' or 'product'

they may be confused about mathematical terms
such as 'odd' or 'table', which have different meanings in everyday English

they may be confused about other words,
like 'area' or 'divide', which are used in everyday English and have similar, though more precise, meanings in mathematics

There are, then, practical reasons why children need to acquire appropriate vocabulary so that they can participate in the activities, lessons and tests that are part of classroom life. There is, however, an even more important reason: mathematical language is crucial to children's development of thinking. If children don't have the vocabulary to talk about division, or perimeters, or numerical difference, they cannot make progress in understanding these areas of mathematical knowledge.

How is the book organised and how can it be used?

To help you introduce appropriate mathematical language at the right time, this book provides four pages of vocabulary checklists for each year group. The first three pages for each year cover mathematical vocabulary relating to the *Framework for Teaching Mathematics*, organised according to its five strands:

numbers and the number system

calculations

solving problems

handling data

measures, shape and space

Using and Applying Mathematics is integrated throughout.

You will also find that sub-headings in this booklet are often based on terms used in the Framework.

The fourth page for each year group lists the language commonly used when giving instructions about mathematical problems, both in questions in national tests and in published resources.

The words listed for each year include vocabulary from the previous year, with new words for the year printed in red from Year 1 onwards.

Class teachers can use these lists to identify the vocabulary relating to a series of lessons they are planning. They can make provision for the introduction of new vocabulary and the consolidation of familiar terms. They can ask support staff and parents to emphasise this vocabulary for an appropriate period.

The checklists are not intended to be exhaustive; you can add more words if you would like to do so.

How do children develop their understanding of mathematical vocabulary?

Teachers often use informal, everyday language in mathematics lessons before or alongside technical mathematical vocabulary. Although this can help children to grasp the meaning of different words and phrases, you will find that a structured approach to the teaching and learning of vocabulary is essential if children are to move on and begin using the correct mathematical terminology as soon as possible.

Some children may start school with a good understanding of mathematical words when used informally, either in English or their home language. Find out the extent of their mathematical vocabulary and the depth of their understanding, and build on this.

You need to plan the introduction of new words in a suitable context, for example, with relevant real objects, mathematical apparatus, pictures and/or diagrams. Explain their meanings carefully and rehearse them several times. Referring to new words only once will do little to promote learning. Encourage their use in context in oral sessions, particularly through your questioning. You can help sort out any ambiguities or misconceptions your pupils may have through a range of open and closed questions. Use every opportunity to draw attention to new words or symbols with the whole class, in a group or when talking to individual pupils. The final stages are learning to read and write new mathematical vocabulary in a range of circumstances, ultimately spelling the relevant words correctly.

Regular, planned opportunities for development

It is not just younger children who need regular, planned opportunities to develop their mathematical vocabulary. All children throughout Key Stages 1 and 2 need to experience a cycle of oral work, reading and writing as outlined below.

oral work based on practical work

so that they have visual images and tactile experience of what mathematical words mean in a variety of contexts

other forms of oral work

so that they have opportunities to:

— listen to adults and other children using the words correctly

— acquire confidence and fluency in speaking, using complete sentences that include the new words and phrases, sometimes in chorus with others and some times individually

— describe, define and compare mathematical properties, positions, methods, patterns, relationships, rules

— discuss ways of tackling a problem, collecting data, organising their work...

— hypothesise or make predictions about possible results

— present, explain and justify their methods, results, solutions or reasoning, to the whole class or to a group or partner

— generalise, or describe examples that match a general statement

reading aloud and silently, sometimes as a whole class and sometimes individually,
for example, reading:

— numbers, signs and symbols, expressions and equations in blackboard presentations

— instructions and explanations in workbooks, textbooks, CD-ROMs...

— texts with mathematical references in fiction and non-fiction books and books of rhymes during the literacy hour as well as mathematics lessons

— labels and captions on classroom displays, in diagrams, graphs, charts and tables...

— definitions in illustrated dictionaries, including dictionaries that they themselves have made, in order to discover synonyms, origins of words, words that start with the same group of letters (such as triangle, tricycle, triplet, trisect...)

writing and recording in a variety of ways, progressing from words, phrases and short sentences to paragraphs and longer pieces of writing,
for example:

— writing prose in order to describe, compare, predict, interpret, explain, justify...

— writing formulae, first using words, then symbols

— sketching and labelling diagrams in order to clarify their meaning

— drawing and labelling graphs, charts or tables, and interpreting and making predictions from the data in them, in mathematics and other subjects

The skill of questioning

Children cannot learn the meanings of words in isolation. The use of questions is crucial in helping them to understand mathematical ideas and use mathematical terms correctly.

It is important to ask questions in different ways so that children who do not understand the first time may pick up the meaning subsequently. Pupils for whom English is an additional language benefit and so will others who are not always familiar with the vocabulary and grammatical structures used in school.

It is easy to use certain types of questions — those that ask the listener to recall and apply facts — more often than those that require a higher level of thinking. If you can use the full range of question types you will find that children begin to give more complex answers in which they explain their thinking.

Types of question

Recalling facts

What is 3 add 7?
How many days are there in a week?
How many centimetres are there in a metre?
Is 31 a prime number?

Applying facts

Tell me two numbers that have a difference of 12.
What unit would you choose to measure the width of the table?
What are the factors of 42?

Hypothesising or predicting

Estimate the number of marbles in this jar.
If we did our survey again on Friday, how likely is it that our graph would be the same?
Roughly, what is 51 times 47?
How many rectangles in the next diagram?
And the next?

Designing and comparing procedures

How might we count this pile of sticks?
How could you subtract 37 from 82?
How could we test a number to see if it is divisible by 6?
How could we find the 20th triangular number?
Are there other ways of doing it?

Interpreting results

So what does that tell us about numbers that end in 5 or 0?
What does the graph tell us about the most common shoe size?
So what can we say about the sum of the angles in a triangle?

Applying reasoning

The seven coins in my purse total 23p. What could they be?
In how many different ways can four children sit at a round table?
Why is the sum of two odd numbers always even?

On this and the following page are further examples of questions to help you promote good dialogue and interaction in mathematics lessons

Below are examples of closed questions with just one correct answer and open questions which have a number of different correct answers. Open questions give more children a chance to respond and they often provide a greater challenge for higher attaining pupils, who can be asked to think of alternative answers and, in suitable cases, to count all the different possibilities.

Closed questions

Count these cubes.

A chew costs 3p. A lolly costs 7p. What do they cost altogether?

What is 6 – 4?

What is 2 + 6 – 3?

Is 16 an even number?

Write a number in each box so that it equals the sum of the two numbers on each side of it.

Copy and complete this addition table.

+	4	7
2		
6		

What are four threes?

What is 7 x 6?

How many centimetres are there in a metre?

Continue this sequence: 1, 2, 4…

What is one fifth add four fifths?

What is 10% of 300?

What is this shape called?

This graph shows room temperature on 19 May.

What was the temperature at 10:00 a.m.?

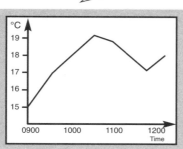

Open questions

How could we count these cubes?

A chew and a lolly cost 10p altogether. What could each sweet cost?

Tell me two numbers with a difference of 2.

What numbers can you make with 2, 3 and 6?

What even numbers lie between 10 and 20?

Write a number in each circle so that the number in each box equals the sum of the two numbers on each side of it. Find different ways of doing it.

Find different ways of completing this table.

	3	4
	7	

Tell me two numbers with a product of 12.

If 7 x 6 = 42, what else can you work out?

Tell me two lengths that together make 1 metre.

Find different ways of continuing this sequence: 1, 2, 4…

Write eight different ways of adding two numbers to make 1.

Find ways of completing: …% of … = 30

Sketch some different triangles.

This graph shows room temperature on 19 May.

Can you explain it?

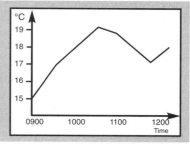

Questions that can help to extend children's thinking

Ask children who are getting started with a piece of work:

How are you going to tackle this?

What information do you have? What do you need to find out or do?

What operation/s are you going to use?

Will you do it mentally, with pencil and paper, using a number line, with a calculator...? Why?

What method are you going to use? Why?

What equipment will you need?

What questions will you need to ask?

How are you going to record what you are doing?

What do you think the answer or result will be? Can you estimate or predict?

Make positive interventions to check progress while children are working, by asking:

Can you explain what you have done so far? What else is there to do?

Why did you decide to use this method or do it this way?

Can you think of another method that might have worked?

Could there be a quicker way of doing this?

What do you mean by...?

What did you notice when...?

Why did you decide to organise your results like that?

Are you beginning to see a pattern or a rule?

Do you think that this would work with other numbers?

Have you thought of all the possibilities? How can you be sure?

Ask children who are stuck:

Can you describe the problem in your own words?

Can you talk me through what you have done so far?

What did you do last time? What is different this time?

Is there something that you already know that might help?

Could you try it with simpler numbers... fewer numbers... using a number line...?

What about putting things in order?

Would a table help, or a picture/diagram/graph?

Why not make a guess and check if it works?

Have you compared your work with anyone else's?

During the plenary session of a lesson ask:

How did you get your answer?

Can you describe your method/pattern/rule to us all? Can you explain why it works?

What could you try next?

Would it work with different numbers?

What if you had started with... rather than...?

What if you could only use...?

Is it a reasonable answer/result? What makes you say so?

How did you check it?

What have you learned or found out today?

If you were doing it again, what would you do differently?

Having done this, when could you use this method/information/idea again?

Did you use any new words today? What do they mean? How do you spell them?

What are the key points or ideas that you need to remember for the next lesson?

Mathematical Vocabulary Checklists

RECEPTION to YEAR 6

RECEPTION

Counting and recognising numbers

COUNTING

number
zero, one, two, three... to twenty and beyond
zero, ten, twenty... one hundred
none
how many...?
count, count (up) to
count on (from, to)
count back (from, to)
count in ones, twos... tens...
more, less
odd, even
every other
how many times?
pattern, pair
guess, estimate
nearly, close to, about the same as
just over, just under
too many, too few, enough, not enough

COMPARING AND ORDERING NUMBERS

the same number as, as many as
Of **two** objects/amounts:
more, larger, bigger, greater
fewer, smaller, less
Of **three** or more objects/amounts:
most, biggest, largest, greatest
fewest, smallest, least
one more, ten more
one less, ten less
compare
order
size
first, second, third... tenth
last, last but one
before, after
next
between

Adding and subtracting

add, more, and
make, sum, total
altogether
score
double
one more, two more, ten more...
how many more to make... ?
how many more is... than...?
take (away), leave
how many are left/left over?
how many have gone?
one less, two less... ten less...
how many fewer is... than...?
difference between
is the same as

Solving problems

REASONING ABOUT NUMBERS OR SHAPES

pattern
puzzle
answer
right, wrong
what could we try next?
how did you work it out?
count, sort
group, set
match
list

PROBLEMS INVOLVING 'REAL LIFE' OR MONEY

compare
double
half, halve
pair
count out, share out
left, left over

money
coin
penny, pence, pound
price
cost
buy
sell
spend, spent
pay
change
dear, costs more
cheap, costs less, cheaper
costs the same as
how much...? how many...?
total

Measures, shape and space

MEASURES (GENERAL)

measure
size
compare
guess, estimate
enough, not enough
too much, too little
too many, too few
nearly, close to, about the same as
just over, just under

LENGTH

length, width, height, depth
long, short, tall
high, low
wide, narrow
deep, shallow
thick, thin
longer, shorter, taller, higher... and so on
longest, shortest, tallest, highest... and so on
far, near, close

MASS

weigh, weighs, balances
heavy/light, heavier/lighter, heaviest/lightest
weight, balance, scales

CAPACITY

full
half full
empty
holds
container

TIME

time
days of the week: Monday, Tuesday...
day, week
birthday, holiday
morning, afternoon, evening, night
bedtime, dinnertime, playtime
today, yesterday, tomorrow
before, after
next, last
now, soon, early, late
quick, quicker, quickest, quickly
slow, slower, slowest, slowly
old, older, oldest
new, newer, newest
takes longer, takes less time
hour, o'clock
clock, watch, hands

RECEPTION

EXPLORING PATTERNS, SHAPE AND SPACE

shape, pattern
flat
curved, straight
round
hollow, solid
corner
face, side, edge, end
sort
make, build, draw

3D SHAPES

cube
pyramid
sphere
cone

2D SHAPES

circle
triangle
square
rectangle
star

PATTERNS AND SYMMETRY

size
bigger, larger, smaller
symmetrical
pattern
repeating pattern
match

POSITION, DIRECTION AND MOVEMENT

position
over, under
above, below
top, bottom, side
on, in
outside, inside
around
in front, behind
front, back
before, after
beside, next to
opposite
apart
between
middle, edge
corner
direction
left, right
up, down
forwards, backwards, sideways

across
next to, close, far
along
through
to, from, towards, away from
movement
slide
roll
turn
stretch, bend

Instructions

listen
join in
say

think
imagine
remember

start from
start with
start at

look at
point to
show me

put, place
arrange
rearrange
change, change over
split
separate

carry on, continue
repeat
what comes next?

find
choose
collect

use
make
build

tell me
describe
pick out
talk about
explain
show me

read
write
trace
copy
complete
finish, end

fill in
shade
colour

tick, cross
draw
draw a line between
join (up)
ring

cost
count
work out
answer
check

GENERAL

same number/s
different number/s
missing number/s
number facts

number line, number track
number square
number cards
counters, cubes, blocks, rods
die, dice
dominoes
pegs, peg board

same way, different way
best way, another way
in order, in a different order

YEAR 1

Numbers and the number system

COUNTING, PROPERTIES OF NUMBERS AND NUMBER SEQUENCES

number
zero, one, two, three... to twenty and beyond
zero, ten, twenty... one hundred
none
how many...?
count, count (up) to
count on (from, to)
count back (from, to)
count in ones, twos... tens...
more, less
odd, even
every other
how many times?
pattern, pair

PLACE VALUE AND ORDERING

units, ones
tens
exchange
digit
'teens' number
the same number as, as many as
equal to
*Of **two** objects/amounts:*
more, larger, bigger, greater
fewer, smaller, less
*Of **three** or more objects/amounts:*
most, biggest, largest, greatest
fewest, smallest, least
one more, ten more
one less, ten less
compare
order
size
first, second, third... tenth, eleventh... twentieth
last, last but one
before, after
next
between, half-way between

ESTIMATING

guess how many, estimate
nearly, roughly, close to
about the same as
just over, just under
too many, too few, enough, not enough

Calculations

ADDITION AND SUBTRACTION

add, more, plus
make, sum, total
altogether
score
double, near double
one more, two more... ten more
how many more to make...?
how many more is... than...?
how much more is...?
subtract, take (away), minus
leave
how many are left/left over? how many are gone?
one less, two less, ten less...
how many fewer is... than...?
how much less is...?
difference between
half, halve
is the same as, equals, sign

Solving problems

MAKING DECISIONS AND REASONING

pattern
puzzle
answer
right, wrong
what could we try next?
how did you work it out?
count out, share out, left, left over
number sentence
sign, operation

MONEY

money
coin
penny, pence, pound
price
cost
buy
sell
spend, spent
pay
change
dear, costs more
cheap, costs less, cheaper
costs the same as
how much...? how many...?
total

Organising and using data

count, sort, vote
list
group, set
list
table

Measures, shape and space

MEASURES (GENERAL)

measure
size
compare
guess, estimate
enough, not enough
too much, too little
too many, too few
nearly, roughly, close to, about the same as
just over, just under

LENGTH

length, width, height, depth
long, short, tall
high, low
wide, narrow
deep, shallow
thick, thin
longer, shorter, taller, higher... and so on
longest, shortest, tallest, highest... and so on
far, near, close
metre
ruler, metre stick

MASS

weigh, weighs, balances
heavy/light, heavier/lighter, heaviest/lightest
weight, balance, scales

CAPACITY

full
half full
empty
holds
container

TIME

time
days of the week: Monday, Tuesday...
seasons: spring, summer, autumn, winter
day, week, month, year,
weekend
birthday, holiday
morning, afternoon, evening

night, midnight
bedtime, dinnertime, playtime
today, yesterday, tomorrow
before, after
next, last
now, soon, early, late
quick, quicker, quickest, quickly
fast, faster, fastest
slow, slower, slowest, slowly
old, older, oldest
new, newer, newest
takes longer, takes less time
hour, o'clock, half past
clock, watch, hands
how long ago?
how long will it be to...?
how long will it take to...?
how often?
always, never, often, sometimes, usually
once, twice

SHAPE AND SPACE

shape, pattern
flat
curved, straight
round
hollow, solid
corner
point, pointed
face, side, edge, end
sort
make, build, draw

3D SHAPES

cube
cuboid
pyramid
sphere
cone
cylinder

2D SHAPES

circle
triangle
square
rectangle
star

PATTERNS AND SYMMETRY

size
bigger, larger, smaller
symmetrical
pattern
repeating pattern
match

Words new to Year 1 are in red

YEAR 1

POSITION, DIRECTION AND MOVEMENT

position
over, under, underneath
above, below
top, bottom, side
on, in
outside, inside
around
in front, behind
front, back
before, after
beside, next to
opposite
apart
between
middle, edge
centre
corner
direction
journey
left, right
up, down
forwards, backwards, sideways
across
next to, close, far
along
through
to, from, towards, away from
movement
slide
roll
turn, whole turn, half turn
stretch, bend

Instructions

listen
join in
say

think
imagine
remember

start from
start with
start at

look at
point to

put, place
arrange
rearrange
change, change over
split
separate

carry on, continue
repeat
what comes next?

find
choose
collect

use
make
build

tell me
describe
pick out
talk about
explain
show me

read
write
record
trace
copy
complete
finish, end

fill in
shade
colour

tick, cross
draw
draw a line between
join (up)
ring
arrow

cost
count
work out
answer
check

GENERAL

same number/s
different number/s
missing number/s
number facts

number line, number track
number square
number cards
abacus
counters, cubes, blocks, rods
die, dice
dominoes
pegs, peg board

same way, different way
best way, another way
in order, in a different order

YEAR 2

Numbers and the number system

COUNTING, PROPERTIES OF NUMBERS AND NUMBER SEQUENCES

number
zero, one, two, three... to twenty and beyond
zero, ten, twenty... one hundred
zero, one hundred, two hundred... one thousand
none
how many...?
count, count (up) to
count on (from, to)
count back (from, to)
count in ones, twos, threes, fours, fives and so on
count in tens
more, less
tally
odd, even
every other
how many times?
multiple of
sequence
continue
predict
pattern, pair, rule

PLACE VALUE AND ORDERING

units, ones
tens, hundreds
digit
one-, two- or three-digit number
'teens' number
place, place value
stands for, represents
exchange
the same number as, as many as
equal to
Of **two** *objects/amounts:*
more, larger, bigger, greater
fewer, smaller, less
Of **three** *or more objects/amounts:*
most, biggest, largest. greatest
fewest, smallest, least
one more, ten more
one less, ten less
compare
order
size
first, second, third... tenth... twentieth
twenty-first, twenty-second...
last, last but one
before, after
next
between, half-way between

ESTIMATING

guess how many, estimate
nearly, roughly, close to
about the same as
just over, just under
exact, exactly
too many, too few, enough, not enough
round, nearest, round to the nearest ten

FRACTIONS

part, equal parts
fraction
one whole
one half, two halves
one quarter, two... three... four quarters

Calculations

ADDITION AND SUBTRACTION

add, addition, more, plus
make, sum, total
altogether
score
double, near double
one more, two more... ten more... one hundred more
how many more to make...?
how many more is... than...?
how much more is...?
subtract, take away, minus
leave, how many are left/left over?
one less, two less... ten less... one hundred less
how many less is... than...?
how much fewer is...?
difference between
half, halve
is the same as, equals, sign
tens boundary

MULTIPLICATION AND DIVISION

lots of, groups of
times, multiply, multiplied by
multiple of
once, twice, three times,
four times, five times... ten times...
times as (big, long, wide and so on)
repeated addition
array
row, column
double, halve
share, share equally
one each, two each, three each...
group in pairs, threes... tens
equal groups of
divide divided by, divided into
left, left over

Solving problems

MAKING DECISIONS AND REASONING

pattern, puzzle
calculate, calculation
mental calculation
jotting
answer
right, correct, wrong
what could we try next?
how did you work it out?
number sentence
sign, operation, symbol

MONEY

money
coin
penny, pence, pound, £
price, cost
buy, bought, sell, sold
spend, spent
pay
change
dear, costs more
cheap, costs less, cheaper
how much...? how many...?
total

Organising and using data

count, tally, sort, vote
graph, block graph, pictogram
represent
group, set
list, table
label, title
most popular, most common
least popular, least common

Measures, shape and space

MEASURES (GENERAL)

measure
size
compare
measuring scale
guess, estimate
enough, not enough
too much, too little
too many, too few
nearly, roughly, about, close to, about the same as
just over, just under

LENGTH

length, width, height, depth
long, short, tall, high, low
wide, narrow, deep, shallow, thick, thin
longer, shorter, taller, higher... and so on
longest, shortest, tallest, highest... and so on
far, further, furthest, near, close
metre, centimetre
ruler, metre stick, tape measure

MASS

weigh, weighs, balances
heavy/light, heavier/lighter, heaviest/lightest
kilogram, half-kilogram, gram
balance, scales, weight

CAPACITY

capacity
full, half full
empty
holds, contains
litre, half-litre, millilitre
container

TIME

time
days of the week: Monday, Tuesday...
months of the year: January, February...
seasons: spring, summer, autumn, winter
day, week, fortnight, month, year
weekend
birthday, holiday
morning, afternoon, evening, night, midnight
bedtime, dinnertime, playtime
today, yesterday, tomorrow
before, after
next, last
now, soon, early, late
quick, quicker, quickest, quickly
fast, faster, fastest
slow, slower, slowest, slowly
old, older, oldest
new, newer, newest
takes longer, takes less time
how long ago?/how long will it be to...?
how long will it take to...?
hour, minute, second
o'clock, half past, quarter to, quarter past
clock, watch, hands
digital/analogue clock/watch, timer
how often?
always, never, often, sometimes, usually
once, twice

YEAR 2

SHAPE AND SPACE

shape, pattern
flat, curved, straight
round
hollow, solid
corner
point, pointed
face, side, edge, end
sort
make, build, draw
surface

3D SHAPES

cube
cuboid
pyramid
sphere
cone
cylinder

2D SHAPES

circle, circular
triangle, triangular
square
rectangle, rectangular
star
pentagon
hexagon
octagon

PATTERNS AND SYMMETRY

size
bigger, larger, smaller
symmetrical
line of symmetry
fold
match
mirror line, reflection
pattern
repeating pattern

POSITION, DIRECTION AND MOVEMENT

position
over, under, underneath
above, below
top, bottom, side
on, in
outside, inside
around
in front, behind
front, back
before, after
beside, next to
opposite
apart
between

middle, edge
centre
corner
direction
journey, route
left, right
up, down
higher, lower
forwards, backwards, sideways
across
next to, close, far
along
through
to, from, towards, away from
clockwise, anti-clockwise
movement
slide
roll
whole turn, half turn, quarter turn
right angle
straight line
clockwise, anti-clockwise
stretch, bend

Instructions

listen
join in
say
recite

think
imagine
remember

start from
start with
start at

look at
point to
show me

put, place
arrange, rearrange
change, change over
split
separate

carry on, continue
repeat
what comes next...?
predict
describe the pattern
describe the rule

find, find all, find different
investigate

choose
decide
collect
use
make
build

tell me
describe
name
pick out
discuss
talk about
explain
explain your method
explain how you got your answer
give an example of...
show how you...

read
write
record
write in figures
present
represent
trace

copy
complete
finish, end

fill in
shade, colour
label

tick, cross
draw
draw a line between
join (up)
ring
arrow

cost, count, tally

calculate
work out
solve
answer
check

GENERAL

same, different
missing number/s
number facts
number pairs
number bonds

number line, number track
number square, hundred square
number cards
number grid
abacus
counters, cubes, blocks, rods
die, dice
dominoes
pegs, peg board
geo-strips

same way, different way
best way, another way
in order, in a different order

Words new to Year 2 are in red

YEAR 3

Numbers and the number system

COUNTING, PROPERTIES OF NUMBERS AND NUMBER SEQUENCES

number
zero, one, two, three... to twenty and beyond
zero, ten, twenty... one hundred
zero, one hundred, two hundred... one thousand
none
how many...?
count, count (up) to
count on (from, to)
count back (from, to)
count in ones, twos, threes, fours, fives and so on
count in tens, hundreds
more, less
tally
odd, even
every other
how many times?
multiple of
sequence
continue
predict
pattern, pair, rule
relationship

PLACE VALUE AND ORDERING

units, ones
tens, hundreds
digit
one-, two- or three-digit number
'teens' number
place, place value
stands for, represents
exchange
the same number as, as many as
equal to
Of **two** *objects/amounts:*
more, larger, bigger, greater
fewer, smaller, less
Of **three** *or more objects/amounts:*
most, biggest, largest, greatest
fewest, smallest, least
one more, ten more, one hundred more
one less, ten less, one hundred less
compare
order
size
first, second, third... tenth... twentieth
twenty-first, twenty-second...
last, last but one
before, after
next
between, half-way between

ESTIMATING

guess how many, estimate
nearly, roughly, close to
approximate, approximately
about the same as
just over, just under
exact, exactly
too many, too few, enough, not enough
round (up or down)
nearest (round to the nearest ten)

FRACTIONS

part, equal parts
fraction
one whole
one half, two halves
one quarter, two... three... four quarters
one third, two thirds
one tenth

Calculations

ADDITION AND SUBTRACTION

add, addition, more, plus
make, sum, total
altogether
score
double, near double
one more, two more... ten more... one hundred more
how many more to make ...?
how many more is... than ...?
how much more is...?
subtract, take (away), minus
leave, how many are left/left over?
one less, two less... ten less... one hundred less
how many fewer is... than ...?
how much less is...?
difference between
half, halve
is the same as, equals, sign
tens boundary, hundreds boundary

MULTIPLICATION AND DIVISION

lots of, groups of
times, product, multiply, multiplied by
multiple of
once, twice, three times,
four times, five times... ten times...
times as (big, long, wide and so on)
repeated addition
array
row, column
double, halve
share, share equally
one each, two each, three each...
group in pairs, threes... tens

equal groups of
divide, divided by, divided into
left, left over, remainder

Solving problems

MAKING DECISIONS AND REASONING

pattern, puzzle
calculate, calculation
mental calculation
method
jotting
answer
right, correct, wrong
what could we try next?
how did you work it out?
number sentence
sign, operation, symbol, equation

MONEY

money
coin, note
penny, pence, pound, £
price, cost
buy, bought, sell, sold
spend, spent
pay
change
dear, costs more, more/most expensive
cheap, costs less, cheaper, less/least expensive
how much…? how many…?
total, amount
value

Handling data

count, tally, sort, vote
graph, block graph, pictogram
represent
group, set
list, chart, bar chart
table, frequency table
Carroll diagram, Venn diagram
label, title, axis, axes
diagram
most popular, most common
least popular, least common

Measures, shape and space

MEASURES (GENERAL)

measure
size
compare
measuring scale, division
guess, estimate
enough, not enough
too much, too little
too many, too few
nearly, roughly, about, close to,
about the same as, approximately
just over, just under

LENGTH

length, width, height, depth
long, short, tall, high, low
wide, narrow, deep, shallow, thick, thin
longer, shorter, taller, higher… and so on
longest, shortest, tallest, highest… and so on
far, further, furthest, near, close
distance apart… between… to… from
kilometre, metre, centimetre
mile
ruler, metre stick, tape measure

MASS

weigh, weighs, balances
heavy/light, heavier/lighter, heaviest/lightest
kilogram, half-kilogram, gram
balance, scales, weight

CAPACITY

capacity
full, half full
empty
holds, contains
litre, half-litre, millilitre
container

TIME

time
days of the week: Monday, Tuesday…
months of the year: January, February…
seasons: spring, summer, autumn, winter
day, week, fortnight, month, year, century
weekend
birthday, holiday
calendar, date
morning, afternoon, evening, night, midnight
a.m., p.m.
bedtime, dinnertime, playtime
today, yesterday, tomorrow

YEAR 3

before, after
next, last
now, soon, early, late, earliest, latest
quick, quicker, quickest, quickly
fast, faster, fastest
slow, slower, slowest, slowly
old, older, oldest
new, newer, newest
takes longer, takes less time
how long ago?/how long will it be to…?
how long will it take to…?
hour, minute, second
o'clock, half past, quarter to, quarter past
clock, watch, hands
digital/analogue clock/watch, timer
how often?
always, never, often, sometimes, usually
once, twice

SHAPE AND SPACE

shape, pattern
flat, curved, straight
round
hollow, solid
corner
point, pointed
face, side, edge, end
sort
make, build, draw
surface
right-angled
vertex, vertices
layer, diagram

3D SHAPES

cube
cuboid
pyramid
sphere, hemi-sphere
cone
cylinder
prism

2D SHAPES

circle, circular, semi-circle
triangle, triangular
square
rectangle, rectangular
star
pentagon, pentagonal
hexagon, hexagonal
octagon, octagonal
quadrilateral

PATTERNS AND SYMMETRY

size
bigger, larger, smaller
symmetrical
line of symmetry
fold
match
mirror line, reflection
pattern
repeating pattern

POSITION, DIRECTION AND MOVEMENT

position
over, under, underneath
above, below
top, bottom, side
on, in
outside, inside
around
in front, behind
front, back
before, after
beside, next to
opposite
apart
between
middle, edge
centre
corner
direction
journey, route, map, plan
left, right
up, down
higher, lower
forwards, backwards, sideways
across
next to, close, far
along
through
to, from, towards, away from
ascend, descend
grid
row, column
clockwise, anti-clockwise
compass point
north, south, east, west, N, S, E, W
horizontal, vertical
diagonal
movement
slide
roll
whole turn, half turn, quarter turn
angle, …is a greater/smaller angle than
right angle
straight line
stretch, bend

Instructions

listen
join in
say
recite

think
imagine
remember

start from
start with
start at

look at
point to
show me

put, place
arrange, rearrange
change, change over
split
separate

carry on, continue
repeat
what comes next?
predict
describe the pattern
describe the rule

find, find all, find different
investigate

choose
decide
collect

use
make
build

tell me
describe
name
pick out
discuss
talk about
explain
explain your method
explain how you got your answer
give an example of...
show how you...
show your working

read
write
record
write in figures

present
represent
interpret
trace
copy
complete
finish, end

fill in
shade, colour
label

tick, cross
draw, sketch
draw a line between
join (up)
ring
arrow

cost, count, tally

calculate
work out
solve
investigate
question
answer
check

GENERAL

same, different
missing number/s
number facts, number pairs, number bonds
greatest value, least value

number line, number track
number square, hundred square
number cards
number grid
abacus
counters, cubes, blocks, rods
die, dice
dominoes
pegs, peg board
geo-strips

same way, different way
best way, another way
in order, in a different order

Words new to Year 3 are in red

YEAR 4

Numbers and the number system

PLACE VALUE, ORDERING AND ROUNDING

units, ones
tens, hundreds, thousands
ten thousand, hundred thousand, million
digit, one-, two-, three- or four-digit number
numeral
'teens' number
place, place value
stands for, represents
exchange
the same number as, as many as
equal to
Of **two** *objects/amounts:*
>, greater than, bigger than, more than, larger than
<, less than, fewer than, smaller than
Of **three** *or more objects/amounts:*
greatest, most, largest, biggest
least, fewest, smallest,
one... ten... one hundred... one thousand more/less
compare, order, size
first... tenth... twentieth
last, last but one
before, after
next
between, half-way between
guess how many, estimate
nearly, roughly, close to, about the same as
approximate, approximately
just over, just under
exact, exactly
too many, too few, enough, not enough
round (up or down), nearest
round to the nearest ten
round to the nearest hundred
integer, positive, negative
above/below zero, minus

PROPERTIES OF NUMBERS AND NUMBER SEQUENCES

number, count, how many...?
odd, even
every other
how many times?
multiple of
digit
next, consecutive
sequence
continue
predict
pattern, pair, rule
relationship
sort, classify, property

FRACTIONS AND DECIMALS

part, equal parts
fraction
one whole
half, quarter, eighth
third, sixth
fifth, tenth, twentieth
proportion, in every, for every
decimal, decimal fraction
decimal point, decimal place

Calculations

ADDITION AND SUBTRACTION

add, addition, more, plus, increase
sum, total, altogether
score
double, near double
how many more to make...?
subtract, take away, minus, decrease
leave, how many are left/left over?
difference between
half, halve
how many more/fewer is... than...?
how much more/less is...?
is the same as, equals, sign
tens boundary, hundreds boundary
inverse

MULTIPLICATION AND DIVISION

lots of, groups of
times, product, multiply, multiplied by
multiple of
once, twice, three times
four times, five times... ten times
times as (big, long, wide, and so on)
repeated addition
array
row, column
double, halve
share, share equally
one each, two each, three each...
group in pairs, threes... tens
equal groups of
divide, divided by, divided into, divisible by
remainder
factor, quotient
inverse

Solving problems

MAKING DECISIONS AND REASONING

pattern, puzzle
calculate, calculation
mental calculation
method
jotting
answer
right, correct, wrong
what could we try next?
how did you work it out?
number sentence
sign, operation, symbol, equation

MONEY

money
coin, note
penny, pence, pound, £
price, cost
buy, bought, sell, sold
spend, spent
pay
change
dear, costs more, more/most expensive
cheap, costs less, cheaper, less/least expensive
how much...? how many...?
total, amount
value

Handling data

count, tally, sort, vote
survey, questionnaire, data
graph, block graph, pictogram
represent
group, set
list, chart, bar chart, tally chart
table, frequency table
Carroll diagram, Venn diagram
label, title, axis, axes
diagram
most popular, most common
least popular, least common

Measures, shape and space

MEASURES (GENERAL)

measure, measurement
size
compare
unit, standard unit
metric unit, imperial unit

measuring scale, division
guess, estimate
enough, not enough
too much, too little
too many, too few
nearly, roughly, about, close to
about the same as, approximately
just over, just under

LENGTH

length, width, height, depth, breadth
long, short, tall, high, low
wide, narrow, deep, shallow, thick, thin
longer, shorter, taller, higher, and so on
longest, shortest, tallest, highest, and so on
far, further, furthest, near, close
distance apart... between... to... from
edge, perimeter
kilometre, metre, centimetre, millimetre
mile
ruler, metre stick, tape measure

MASS

mass: big, bigger, small, smaller, balances
weight: heavy/light, heavier/lighter, heaviest/lightest
weigh, weighs
kilogram, half-kilogram, gram
balance, scales

CAPACITY

capacity
full, half full
empty
holds, contains
litre, half-litre, millilitre
pint
container, measuring cylinder

AREA

area, covers, surface
square centimetre (cm²)

TIME

time
days of the week: Monday, Tuesday...
months of the year: January, February...
seasons: spring, summer, autumn, winter
day, week, fortnight, month
year, leap year, century, millennium
weekend
birthday, holiday
calendar, date, date of birth
morning, afternoon, evening, night

Words new to Year 4 are in red

YEAR 4

a.m., p.m., noon, midnight
today, yesterday, tomorrow
before, after, next, last
now, soon, early, late, earliest, latest
quick, quicker, quickest, quickly
fast, faster, fastest, slow, slower, slowest, slowly
old, older, oldest, new, newer, newest
takes longer, takes less time
how long ago?/how long will it be to...?
how long will it take to...?
timetable, arrive, depart
hour, minute, second
o'clock, half past, quarter to, quarter past
clock, watch, hands
digital/analogue clock/watch, timer
how often?
always, never, often, sometimes, usually

SHAPE AND SPACE

shape, pattern
flat, line
curved, straight
round
hollow, solid
corner
point, pointed
face, side, edge, end
sort
make, build, construct, draw, sketch
centre, radius, diameter
net
surface
angle, right-angled
base, square-based
vertex, vertices
layer, diagram
regular, irregular
concave, convex
open, closed

3D SHAPES

3D, three-dimensional
cube
cuboid
pyramid
sphere, hemi-sphere, spherical
cone
cylinder, cylindrical
prism
tetrahedron, polyhedron

2D SHAPES

2D, two-dimensional
circle, circular, semi-circle
triangle, triangular

equilateral triangle, isosceles triangle
square
rectangle, rectangular, oblong
pentagon, pentagonal
hexagon, hexagonal
heptagon
octagon, octagonal
polygon
quadrilateral

PATTERNS AND SYMMETRY

size
bigger, larger, smaller
symmetrical
line of symmetry, line symmetry
fold
match
mirror line, reflection, reflect, translation
pattern, repeating pattern

POSITION, DIRECTION AND MOVEMENT

position
over, under, underneath
above, below, top, bottom, side
on, in, outside, inside, around
in front, behind, front, back
before, after, beside, next to
opposite, apart
between, middle, edge, centre
corner
direction
journey, route, map, plan
left, right
up, down, higher, lower
forwards, backwards, sideways, across
next to, close, far
along, through, to, from, towards, away from
ascend, descend
grid
row, column
origin, coordinates
clockwise, anti-clockwise
compass point, north, south, east, west, N, S, E, W
north-east, north-west, south-east, south-west
NE, NW, SE, SW
horizontal, vertical, diagonal
movement
slide, roll
whole turn, half turn, quarter turn, rotate
angle, ...is a greater/smaller angle than
right angle
degree
straight line
stretch, bend
ruler, set square
angle measurer, compasses

Instructions

listen, join in, say, recite
think, imagine, remember
start from, start with, start at
look at, point to, show me

put, place
arrange, rearrange
change, change over
split, separate

carry on, continue, repeat
what comes next? predict
describe the pattern, describe the rule

find, find all, find different
investigate

choose, decide
collect

use, make, build, construct

tell me, describe, name, pick out
discuss, talk about
explain
explain your method
explain how you got your answer
give an example of...
show how you...
show your working
justify
make a statement

read, write, record
write in figures
present, represent
interpret
trace, copy
complete, finish, end

fill in, shade, colour
label, plot

tick, cross
draw, sketch
draw a line between, join (up), ring, arrow

cost, count, tally
calculate, work out, solve
investigate, question
answer
check

GENERAL

same, different
missing number/s
number facts, number pairs, number bonds
greatest value, least value

number line, number track
number square, hundred square
number cards, number grid
abacus
counters, cubes, blocks, rods
die, dice
dominoes
pegs, peg board, pin board
geo-strips

same way, different way
best way, another way
in order, in a different order

Words new to Year 4 are in red

YEAR 5

Numbers and the number system

PLACE VALUE, ORDERING AND ROUNDING

units, ones
tens, hundreds, thousands
ten thousand, hundred thousand, million
digit, one-, two-, three- or four-digit number
numeral
'teens' number
place, place value
stands for, represents
exchange
the same number as, as many as
equal to
Of **two** *objects/amounts:*
>, greater than, more than, larger than, bigger than
<, less than,fewer than, smaller than
≥, greater than or equal to
≤, less than or equal to
Of **three** *or more objects/amounts:*
greatest, most, largest, biggest
least, fewest, smallest
one... ten... one hundred... one thousand more/less
compare, order, size
ascending/descending order
first... tenth... twentieth
last, last but one
before, after, next
between, half-way between
guess how many, estimate
nearly, roughly, close to, about the same as
approximate, approximately
≈, is approximately equal to
just over, just under
exact, exactly
too many, too few, enough, not enough
round (up or down), nearest
round to the nearest ten/hundred
round to the nearest thousand
integer
positive, negative
above/below zero, minus

PROPERTIES OF NUMBERS AND NUMBER SEQUENCES

number, count, how many...?
odd, even
every other
how many times?
multiple of
digit
next, consecutive
sequence
continue
predict
pattern, pair, rule

relationship
sort, classify, property
formula
divisible (by), divisibility, factor
square number

FRACTIONS, DECIMALS, PERCENTAGES, RATIO AND PROPORTION

part, equal parts
fraction, proper/improper fraction
mixed number
numerator, denominator
equivalent, reduced to, cancel
one whole
half, quarter, eighth
third, sixth, ninth, twelfth
fifth, tenth, twentieth, hundredth
proportion, in every, for every
to every, as many as
decimal, decimal fraction
decimal point, decimal place
percentage, per cent, %

Calculations

ADDITION AND SUBTRACTION

add, addition, more, plus, increase
sum, total, altogether
score
double, near double
how many more to make...?
subtract, take away, minus, decrease
leave, how many are left/left over?
difference between
half, halve
how many more/ fewer is... than...?
how much more/less is...?
is the same as, equals, sign
tens boundary, hundreds boundary
units boundary, tenths boundary
inverse

MULTIPLICATION AND DIVISION

lots of, groups of
times, product, multiply, multiplied by
multiple of
once, twice, three times
four times, five times... ten times
times as (big, long, wide, and so on)
repeated addition
array
row, column
double, halve
share, share equally

one each, two each, three each...
group in pairs, threes... tens
equal groups of
divide, divided by, divided into, divisible by
remainder
factor, quotient
inverse

USING A CALCULATOR

calculator
display, key, enter, clear
constant

Solving problems

MAKING DECISIONS AND REASONING

pattern, puzzle
calculate, calculation
mental calculation
method, strategy
jotting
answer
right, correct, wrong
what could we try next?
how did you work it out?
number sentence
sign, operation, symbol, equation

MONEY

money
coin, note
penny, pence, pound, £
price, cost
buy, bought, sell, sold
spend, spent
pay
change
dear, costs more, more/most expensive
cheap, costs less, cheaper, less/least expensive
how much...? how many...?
total, amount, value
discount
currency

Handling data

count, tally, sort, vote
survey, questionnaire
data, database
graph, block graph, line graph
pictogram,
represent
group, set
list, chart, bar chart, bar line chart
tally chart
table, frequency table

Carroll diagram, Venn diagram
label, title, axis, axes
diagram
most popular, most common
least popular, least common
mode, range
maximum/minimum value
classify, outcome

PROBABILITY

fair, unfair
likely, unlikely, likelihood
certain, uncertain
probable, possible, impossible
chance, good chance
poor chance, no chance
risk, doubt

Measures, shape and space

MEASURES (GENERAL)

measure, measurement
size
compare
unit, standard unit
metric unit, imperial unit
measuring scale, division
guess, estimate
enough, not enough
too much, too little
too many, too few
nearly, roughly, about, close to
about the same as, approximately
just over, just under

LENGTH

length, width, height, depth, breadth
long, short, tall, high, low
wide, narrow, deep, shallow, thick, thin
longer, shorter, taller, higher... and so on
longest, shortest, tallest, highest... and so on
far, further, furthest, near, close
distance apart... between... to... from
edge, perimeter
kilometre, metre, centimetre, millimetre
mile
ruler, metre stick, tape measure

MASS

mass: big, bigger, small, smaller, balances
weight: heavy/light, heavier/lighter, heaviest/lightest
weigh, weighs
kilogram, half-kilogram, gram
balance, scales

YEAR 5

CAPACITY

capacity
full, half full
empty
holds, contains
litre, half-litre, millilitre
pint, gallon
container, measuring cylinder

AREA

area, covers, surface
square centimetre (cm^2), square metre (m^2)
square millimetre (mm^2)

TIME

time
days of the week: Monday, Tuesday...
months of the year: January, February...
seasons: spring, summer, autumn, winter
day, week, fortnight, month
year, leap year, century, millennium
weekend
birthday, holiday
calendar, date, date of birth
morning, afternoon, evening, night
a.m., p.m., noon, midnight
today, yesterday, tomorrow
before, after, next, last
now, soon, early, late, earliest, latest
quick, quicker, quickest, quickly
fast, faster, fastest, slow, slower, slowest, slowly
old, older, oldest, new, newer, newest
takes longer, takes less time
how long ago?/how long will it be to…?
how long will it take to...?
timetable, arrive, depart
hour, minute, second
o'clock, half past, quarter to, quarter past
clock, watch, hands
digital/analogue clock/watch, timer
24-hour clock, 12-hour clock
how often?
always, never, often, sometimes, usually

SHAPE AND SPACE

shape, pattern
flat, line
curved, straight
round
hollow, solid
corner
point, pointed
face, side, edge, end
sort
make, build, construct, draw, sketch
centre, radius, diameter

net
surface
angle, right-angled
congruent
base, square-based
vertex, vertices
layer, diagram
regular, irregular
concave, convex
open, closed

3D SHAPES

3D, three-dimensional
cube, cuboid
pyramid
sphere, hemi-sphere, spherical
cone
cylinder, cylindrical
prism
tetrahedron, polyhedron, octahedron

2D SHAPES

2D, two-dimensional
circle, circular, semi-circle
triangle, triangular
equilateral triangle, isosceles triangle, scalene triangle
square
rectangle, rectangular, oblong
pentagon, pentagonal
hexagon, hexagonal
heptagon
octagon, octagonal
polygon
quadrilateral

PATTERNS AND SYMMETRY

size
bigger, larger, smaller
symmetrical
line of symmetry, axis of symmetry
line symmetry, reflective symmetry
fold
match
mirror line, reflection, reflect, translation
pattern, repeating pattern

POSITION, DIRECTION AND MOVEMENT

position
over, under, underneath
above, below, top, bottom, side
on, in, outside, inside, around
in front, behind, front, back
before, after, beside, next to
opposite

apart
between, middle, edge, centre
corner
direction
journey, route, map, plan
left, right
up, down, higher, lower
forwards, backwards, sideways, across
next to, close, far
along, through, to, from, towards, away from
ascend, descend
grid, row, column
origin, coordinates
clockwise, anti-clockwise
compass point, north, south, east, west, N, S, E, W
north-east, north-west, south-east, south-west
NE, NW, SE, SW
horizontal, vertical, diagonal
parallel, perpendicular
x-axis, y-axis
quadrant
movement
slide, roll
whole turn, half turn, quarter turn
rotate, rotation
angle, ...is a greater/smaller angle than
right angle, acute, obtuse
degree
straight line
stretch, bend
ruler, set square
angle measurer, compasses, protractor

Instructions

listen, join in, say, recite
think, imagine, remember
start from, start with, start at
look at, point to, show me

put, place
arrange, rearrange
change, change over
split, separate

carry on, continue, repeat
what comes next?, predict
describe the pattern, describe the rule

find, find all, find different
investigate

choose, decide
collect

use, make, build, construct, bisect

tell me, describe, name, pick out, identify
discuss, talk about

explain
explain your method/answer/reasoning
give an example of...
show how you...
show your working
justify
make a statement

read, write, record
write in figures
present, represent
interpret
trace, copy
complete, finish, end

fill in, shade, colour
label, plot

tick, cross
draw, sketch
draw a line between, join (up), ring, arrow

cost, count, tally

calculate, work out, solve, convert
investigate, question
answer
check

GENERAL

same, different
missing number/s
number facts, number pairs, number bonds
greatest value, least value

number line, number track
number square, hundred square
number cards, number grid
abacus
counters, cubes, blocks, rods
die, dice, spinner
dominoes
pegs, peg board, pin board
geo-strips

same way, different way
best way, another way
in order, in a different order

Words new to Year 5 are in red

YEAR 6

Numbers and the number system

PLACE VALUE, ORDERING AND ROUNDING

units, ones
tens, hundreds, thousands
ten thousand, hundred thousand, million
digit, one-, two-, three- or four-digit number
numeral
'teens' number
place, place value
stands for, represents
exchange
the same number as, as many as
equal to
Of **two** *objects/amounts:*
>, greater than, more than, larger than, bigger than
<, less than, fewer than, smaller than
≥, greater than or equal to
≤, less than or equal to
Of **three** *or more objects/amounts:*
greatest, most, largest, biggest
least, fewest, smallest,
one... ten... one hundred... one thousand more/less
compare, order, size
ascending/descending order
first... tenth... twentieth
last, last but one
before, after
next
between, half-way between
guess how many, estimate
nearly, roughly, close to, about the same as
approximate, approximately
≈, is approximately equal to
just over, just under
exact, exactly
too many, too few, enough, not enough
round (up or down), nearest
round to the nearest ten/hundred/thousand
integer, positive, negative
above/below zero, minus

PROPERTIES OF NUMBERS AND NUMBER SEQUENCES

number, count, how many...?
odd, even
every other
how many times?
multiple of
digit
next, consecutive
sequence
continue
predict
pattern, pair, rule

relationship
sort, classify, property
formula
divisible (by), divisibility, factor, factorise
square number
prime, prime factor

FRACTIONS, DECIMALS, PERCENTAGES, RATIO AND PROPORTION

part, equal parts
fraction, proper/improper fraction
mixed number
numerator, denominator
equivalent, reduced to, cancel
one whole
half, quarter, eighth
third, sixth, ninth, twelfth
fifth, tenth, twentieth
hundredth, thousandth
proportion, in every, for every
to every, as many as
decimal, decimal fraction
decimal point, decimal place
percentage, per cent, %

Calculations

ADDITION AND SUBTRACTION

add, addition, more, plus, increase
sum, total, altogether
score
double, near double
how many more to make...?
subtract, take away, minus, decrease
leave, how many are left/left over?
difference between
half, halve
how many more/fewer is... than...?
how much more/less is...?
is the same as, equals, sign
tens boundary, hundreds boundary
units boundary, tenths boundary
inverse

MULTIPLICATION AND DIVISION

lots of, groups of
times, product, multiply, multiplied by
multiple of
once, twice, three times
four times, five times... ten times
times as (big, long, wide, and so on)
repeated addition
array, row, column
double, halve
share, share equally

one each, two each, three each...
group in pairs, threes... tens
equal groups of
divide, divided by, divided into, divisible by
remainder
factor, quotient
inverse

USING A CALCULATOR

calculator, display, key
enter, clear, sign change
constant, recurring, memory, operation key

Solving problems

MAKING DECISIONS AND REASONING

pattern, puzzle
calculate, calculation
mental calculation
method, strategy
jotting
answer
right, correct, wrong
what could we try next?
how did you work it out?
number sentence
sign, operation, symbol, equation

MONEY

money
coin, note
penny, pence, pound, £
price, cost
buy, bought, sell, sold
spend, spent
pay
change
dear, costs more, more/most expensive
cheap, costs less, cheaper, less/least expensive
how much...? how many...?
total, amount, value
discount, profit, loss
currency

Handling data

count, tally, sort, vote
survey, questionnaire
data, database
graph, block graph, line graph
pictogram,
represent
group, set
list, chart, bar chart, bar line chart
tally chart

table, frequency table
Carroll diagram, Venn diagram
label, title, axis, axes
diagram
most popular, most common
least popular, least common
mode, range, mean, average, median
statistics, distribution
maximum/minimum value
classify, outcome

PROBABILITY

fair, unfair
likely, unlikely, likelihood, equally likely
certain, uncertain
probable, possible, impossible
chance, good chance,
poor chance, no chance
equal chance, even chance, fifty-fifty chance
risk, doubt
biased, random

Measures, shape and space

MEASURES (GENERAL)

measure, measurement
size
compare
unit, standard unit
metric unit, imperial unit
measuring scale, division
guess, estimate
enough, not enough
too much, too little
too many, too few
nearly, roughly, about, close to
about the same as, approximately
just over, just under

LENGTH

length, width, height, depth, breadth
long, short, tall, high, low
wide, narrow, deep, shallow, thick, thin
longer, shorter, taller, higher, and so on
longest, shortest, tallest, highest, and so on
far, further, furthest, near, close
distance apart... between... to... from
edge, perimeter, circumference
kilometre, metre, centimetre, millimetre
mile, yard, feet, foot, inches, inch
ruler, metre stick, tape measure, compasses

MASS

mass: big, bigger, small, smaller, balances
weight: heavy/light, heavier/lighter, heaviest/lightest
weigh, weighs

YEAR 6

tonne, kilogram, half-kilogram, gram
pound, ounce
balance, scales

CAPACITY

capacity
full, half full
empty
holds, contains
litre, half-litre, centilitre, millilitre
pint, gallon
container, measuring cylinder

AREA

area, covers, surface
square centimetre (cm²), square metre (m²)
square millimetre (mm²)

TIME

time
days of the week: Monday, Tuesday...
months of the year: January, February...
seasons: spring, summer, autumn, winter
day, week, fortnight, month
year, leap year, century, millennium
weekend
birthday, holiday
calendar, date, date of birth
morning, afternoon, evening, night
a.m., p.m., noon, midnight
today, yesterday, tomorrow
before, after, next, last
now, soon, early, late, earliest, latest
quick, quicker, quickest, quickly
fast, faster, fastest, slow, slower, slowest, slowly
old, older, oldest, new, newer, newest
takes longer, takes less time
how long ago?/how long will it be to...?
how long will it take to...?
timetable, arrive, depart
hour, minute, second
o'clock, half past, quarter to, quarter past
clock, watch, hands
digital/analogue clock/watch, timer
24-hour clock, 12-hour clock
Greenwich Mean Time, British Summer Time
International Date Line
how often?
always, never, often, sometimes, usually

SHAPE AND SPACE

shape, pattern
flat, line
curved, straight
round
hollow, solid
corner

point, pointed
face, side, edge, end
sort
make, build, construct, draw, sketch
centre, radius, diameter
circumference, concentric, arc
net
surface
angle, right-angled
congruent
intersecting, intersection
plane
base, square-based
vertex, vertices
layer, diagram
regular, irregular
concave, convex
open, closed
tangram

3D SHAPES

3D, three-dimensional
cube, cuboid
pyramid
sphere, hemi-sphere, spherical
cone
cylinder, cylindrical
prism
tetrahedron, polyhedron, octahedron, dodecahedron

2D SHAPES

2D, two-dimensional
circle, circular, semi-circle
triangle, triangular
equilateral triangle, isosceles triangle, scalene triangle
square, rhombus
rectangle, rectangular, oblong
pentagon, pentagonal
hexagon, hexagonal
heptagon
octagon, octagonal
polygon
quadrilateral
kite
parallelogram, trapezium

PATTERNS AND SYMMETRY

size
bigger, larger, smaller
symmetrical
line of symmetry, axis of symmetry
line symmetry, reflective symmetry
fold
match
mirror line, reflection, reflect, translation
pattern, repeating pattern

POSITION, DIRECTION AND MOVEMENT

position
over, under, underneath
above, below, top, bottom, side
on, in, outside, inside, around
in front, behind, front, back
before, after, beside, next to
opposite, apart
between, middle, edge, centre
corner
direction
journey, route, map, plan
left, right
up, down, higher, lower
forwards, backwards, sideways, across
next to, close, far
along, through, to, from, towards, away from
ascend, descend
grid, row, column
origin, coordinates
clockwise, anti-clockwise
compass point, north, south, east, west, N, S, E, W
north-east, north-west, south-east, south-west
NE, NW, SE, SW
horizontal, vertical, diagonal
parallel, perpendicular
x-axis, y-axis
quadrant
movement
slide, roll
whole turn, half turn, quarter turn, rotate, rotation
angle, ...is a greater/smaller angle than
right angle, acute, obtuse, reflex
degree
straight line
stretch, bend
ruler, set square
angle measurer, compasses, protractor

Instructions

listen, join in, say, recite
think, imagine, remember
start from, start with, start at
look at, point to, show me

put, place
arrange, rearrange
change, change over
adjusting, adjust
split, separate

carry on, continue, repeat
what comes next?, predict
describe the pattern, describe the rule

find, find all, find different
investigate

choose, decide
collect
use, make, build, construct, bisect

tell me, define, describe, name, pick out, identify
discuss, talk about
explain
explain your method/answer/reasoning
give an example of...
show how you...
show your working
justify
make a statement

read, write, record
write in figures
present, represent
interpret
trace, copy
complete, finish, end

fill in, shade, colour
label, plot

tick, cross
draw, sketch
draw a line between, join (up), ring, arrow

cost, count, tally

calculate, work out, solve, convert

investigate, interrogate (data), question, prove
answer
check

GENERAL

same, identical, different
missing number/s
number facts, number pairs, number bonds
greatest value, least value

number line, number track
number square, hundred square
number cards, number grid
abacus
counters, cubes, blocks, rods
die, dice, spinner
dominoes
pegs, peg board, pin board
geo-strips

same way, different way
best way, another way
in order, in a different order

Words new to Year 6 are in red

Mathematical dictionaries

Every classroom needs a mathematical dictionary, suited to the age of the children. This could either be a published version, or one which the children have made themselves. As well as being useful for children to look up the meanings of words, it will be on hand when the teacher needs to refer to a mathematical dictionary.